Cesar Barros Ferreira
Joel Medeiros Bezerra

Environmental impacts of a red ceramics industry in Barcelona/RN

Cesar Barros Ferreira
Joel Medeiros Bezerra

Environmental impacts of a red ceramics industry in Barcelona/RN

Exploratory survey

ScienciaScripts

Imprint

Cover image: www.ingimage.com

This book is a translation from the original published under ISBN 978-3-330-76718-8.

Publisher:
Sciencia Scripts
is a trademark of
Dodo Books Indian Ocean Ltd. and OmniScriptum S.R.L publishing group

120 High Road, East Finchley, London, N2 9ED, United Kingdom
Str. Armeneasca 28/1, office 1, Chisinau MD-2012, Republic of Moldova, Europe
Managing Directors: Ieva Konstantinova, Victoria Ursu
info@omniscriptum.com

Printed at: see last page
ISBN: 978-620-8-56696-8

"The cleanest environment is not the one that gets cleaned the most, but the one that gets dirty the least."

Francisco Cândido Xavier (Chico Xavier)

SUMMARY

Environmental preservation has been one of the most discussed issues in recent decades, as the rate of pollution has increased. In view of this observation, this study seeks to assess the environmental impacts caused by a ceramics industry in the city of Barcelona, in the state of Rio Grande do Norte. In order to do this, it was necessary to diagnose the situation of the company being researched, identify and measure the potential and actual environmental impacts caused throughout the process, extraction and production of the final product. In order to carry out this research, a case study was carried out on the company analysed. The data was assessed quantitatively using the Leopold Matrix, defining the positive and negative impacts of the entire production process in the ceramics industry. As a result, several negative impacts were observed from the moment the clay was extracted, generating erosion and other transformations in the environment, as well as during the firing process, through the burning of firewood and the emission of gases harmful to the ozone layer. It was also found that the sale of the wood used in the production process should also be monitored, as some suppliers do not have an environmental licence to extract it. This type of activity needs to adopt environmental management tools integrated into the production process in order to develop and expand this activity in a sustainable way.

Key words: Environment. Environmental impact. Ceramics industry.

SUMMARY

CHAPTER 1

INTRODUCTION

Environmental preservation has been one of the most discussed issues in recent decades, given the major negative effects caused by the lack of control and management of degrading anthropic activities, as well as the deficiency in environmental preservation on the part of man. In the current world scenario, there is devastation of forests, pollution of rivers, lack of basic sanitation and sewage treatment, improper disposal of solid and liquid waste, which are just some of the causes of the negative environmental impacts that can be observed, as well as the emergence of diseases and sudden climate changes, causing major floods in some areas and droughts, and their respective impacts.

Even though this is a phenomenon that occurs in major industrial hubs and urban centres, the industrial activities of some small towns also contribute to the increase in these negative impacts, which are often still invisible to the population living around these small and medium-sized industries, since their socio-economic contribution speaks louder and the consequences can only be seen in the medium and long term, with the emergence of diseases, the accumulation of irregularly disposed waste, the inversion of climatic conditions, etc.

The ceramics industry is of significant economic importance in the area in which it operates. The economic growth of the cities in which the ceramics industry operates can be seen in the positive impact it has had on job creation, income and the expansion of urban infrastructure. This leads to a better quality of life for the population.

The industrial sector responsible for ceramics production, the main focus of this study, is exempt from some of the most common types of environmental pollution in our country, such as river, lake and ocean pollution and groundwater pollution. However, deforestation is necessary for the firing of its finished products, which also generates carbon dioxide emissions into the atmosphere, and possibly the wear and tear of deforested environments and their respective alteration of the landscape and the ecological niche present, as well as medium and long-term impacts.

1.1 RESEARCH OBJECTIVES

In order for this research to achieve its aims, it is therefore necessary to list the objectives that will serve as guiding elements throughout the development of this work and which will be presented in this section.

1.1.1 General objective

To assess the environmental impacts associated with the production activities that a ceramics industry could cause in the municipality of Barcelona-RN.

1.1.2 Specific objectives

- Environmental diagnosis of the situation of a ceramics industry and its main activities in the municipality of Barcelona-RN;
- Identify the potential and actual environmental impacts caused during and after the production process in the ceramics industry;
- Evaluate the environmental impacts during each stage of the ceramics production process in the municipality of Barcelona-RN, using the Leopold matrix.

CHAPTER 2

THEORETICAL FRAMEWORK

2.1 THE CERAMICS INDUSTRY IN BRAZIL

Researchers confirm that ceramics is the oldest of all industries, since it was born when man began to use fire-hardened clay, perfecting the hardening process that was obtained by chance and multiplying its use. As time went by, ceramics began to replace worked stone, wood and even household utensils made from fruits such as coconuts or the shells of certain cucurbits (BYLAARDT et al, 2015).

Ceramics or ceramic material can be defined as artificial stone obtained by moulding, drying and firing clays or clay mixtures. However, in some cases, some of the stages of this process can be omitted without the necessary raw material ceasing to be clay (LEGGERINI, 2010).

Red ceramics include reddish-coloured materials used in construction, such as bricks, blocks, tiles, hollow elements, slabs, ceramic tubes and expanded clays, as well as household utensils and decorative items. Slabs, on the other hand, fall within this classification. However, they should be considered as cladding materials (ABCERAM, 2011).

The ceramics tradition is traced back to the time of the Discovery and that it was the Portuguese and their slaves who were mainly responsible for this tradition in our country. However, archaeology debunks this farce by revealing that the Indians had already mastered the technique of working with clay and even ceramics, especially on the island of Marajó in the state of Pará.

According to the records of the National Association of Manufacturers of Ceramics for Coatings, Sanitary Ware and Congeners (ANFACER), ceramics in Brazil had its beginnings on the island of Marajó as an advanced indigenous culture. However, in various parts of the Amazon region, there are signs of simpler ceramics, dating back around five thousand years.

The pottery produced by the indigenous people on Marajó Island was highly elaborate and used various techniques, such as scraping, incision, excision and painting. The modelling used is typically anthropomorphic, although there are some examples produced with reliefs in the shape of snakes and lizards. Among the ceramic objects found on the island, the following stand out: stools, figurines, spindle whorls, thongs, spoons, ear and lip adornments, whistles and miniature vases. It's worth pointing out that even though they were unaware of the lathe and used rudimentary tools, the ceramics produced by the indigenous people of this region are considered to be of high value, which indicates that they surpassed the primitive stages of the Stone and Bronze Ages, respectively (ANFANCER, 2015).

The ceramics sector in Brazil is divided into segments, which differ in terms of the products they manufacture and the markets in which they operate. The main segments are: Structural Ceramics (Red); Coverings (floors and tiles); Natural Raw Materials; Refractories; Technical Ceramics and Special Ceramics; Sanitary Ware; Tableware and Adornment, Frits, Glazes and Dyes; Synthetic Raw Materials; Electrical Ceramics; Equipment for Ceramics; and Abrasives (BUSTAMANTE; BRESSIANI, 2000).

With regard to red ceramics, the first signs of its use, there is only some inaccurate information about its use during the Colonial period, based on rudimentary production techniques introduced by the Jesuits when they needed bricks to build colleges and convents. The production of building materials began to be stimulated with the arrival of Tomé de Souza in 1549, which was necessary for the development of better planned and designed cities. However, the development of ceramics began to take place with greater commitment from 1575 onwards, as the manufacture of roof tiles was needed for the formation of the village that would eventually become the city of São Paulo/SP, with the potteries being the initial landmark of São Paulo's industry (SEBRAE, 2008).

But ceramics doesn't just stand out for its pottery industry. The tradition of ceramics in Brazil is maintained by many artisans spread across different states of the country. In the Northeast, for example, Mestre Vitalino's production is very famous, as he and his followers represent scenes and customs of the people with clay figures. In the interior of the state of São Paulo, more precisely in the Paraíba Valley, very beautiful ceramics are made, painted in white and red, and in Minas Gerais, in the Jequitinhonha Valley, light clay is used in their handicraft production (POINT DA ARTE, 2011).

With regard to the ceramic tile industry in Brazil, its production is concentrated between Santa Catarina, São Paulo and a few cities in the north-east of Brazil. The region of Criciúma, in Santa Catarina, is now recognised as an international hub, as it concentrates the largest number of national companies in this segment, producing with wet technology and competing with each other on design and brand, in higher price ranges. In São Paulo, ceramics production is concentrated in two centres: Mogi Guaçu, where companies work with wet technology, while in Santa Gertrudes most companies use dry technology. The Northeast region is in an expansion phase and could soon become an important industrial hub for this segment, given the favourable conditions of raw materials, viable energy and a developing potential consumer market, as well as its good geographical location for exports (AFANCER, 2015).

The ceramics sector is undergoing extensive growth and development, and at the same time it is a very competitive and fierce market segment. In terms of the number of existing companies, the majority are in São Paulo (25.23%), followed by Minas Gerais (12.85%) and Rio Grande do Sul (11.44%). Santa Catarina (9.52%) and Paraná (9.06%) are also found in this context. In the Northeast, the concentration and distribution of the ceramics industry is mainly in the states of Ceará, Bahia and Pernambuco, followed by Rio Grande do Norte, Maranhão and Piauí (SEBRAE, 2014).

2.2 THE RED CERAMICS INDUSTRY IN THE STATE OF RIO GRANDE DO NORTE

According to the document entitled "O perfil da Cerâmica Vermelha do Rio Grande do Norte: uma síntese" (The profile of red ceramics in Rio Grande do Norte: a synthesis), in 2000 there were 206 ceramics factories in the state, the majority of which are in full production, only a few of which have been shut down or temporarily deactivated due to physiographic conditions or financial reasons. The active ceramics factories are spread across 53 municipalities in the state of Rio Grande do Norte, producing roof tiles, bricks and flagstones (SILVA, 2007).

The companies in this segment are predominantly located in rural areas, concentrated in regions close to Natal, in the Açu Valley and in the Seridó region. In the Vale do Açu micro-region, for example, made up of the municipalities of Alto dos Rodrigues, Carnaubais, Itajá, Jucurutu, Pendências, Porto do Mangue and São Rafael, the state's red ceramics industry had the second largest inland production of tiles in 2000, with 46 companies responsible for producing more than 18 million tiles per month, with a predominant production of roof tiles (53%) as well as bricks and flagstones. In this microregion, the municipality of Itajá/RN stands out as having the largest number of red ceramics companies in the Açu Valley microregion, with 16 active enterprises, representing an average of 35% of the total number of companies in the entire microregion, where the main economic activity is precisely the manufacture of products from the red ceramics industry (FIERN/SENAIS, 2001).

Based on data collected from research into red ceramics in the state of Rio Grande do Norte, it can be seen that the development of the ceramics industry takes place through a very complex production process, involving some specific phases such as: extraction of the raw material (be it clay or red clay), mixing or moulding, drying or firing, as well as the final destination of the finished products. As they still use very archaic techniques in their production process, there is a need to consume firewood in the process of firing the products and the inability to reduce production losses, improve productivity and the quality of the products produced. Despite these shortcomings, the ceramics sector has signalled new potential, as it is present throughout the state and generates a number of jobs, income and the social reproduction of part of society in northern Rio Grande do Sul (SILVA, 2007).

According to Silva (2007), the state of Rio Grande do Norte produces 50,186 million tiles per month, which is more than the state of Minas Gerais, which produces around 45 million tiles per month. Thus, in the ceramics sector, Rio Grande do Norte is the largest producer of roof tiles in Brazil, with the Seridó region standing out, which is responsible for almost 95% of the production destined for other states in the Northeast, especially Bahia, Pernambuco, Alagoas and Sergipe.

However, these activities are liable to generate different environmental impacts due to the acquisition of raw materials, processing and consequent generation of waste.

2.3 ENVIRONMENTAL MANAGEMENT

In order to discuss the issue of environmental management, it is *first* necessary *to* identify what the term "environment" means for this area of study. In the field of environmental impact assessment, this term defines the scope of environmental studies, mitigating or compensatory measures, and environmental management plans and programmes. This definition, in this sense, is decisive in defining the scope of environmental planning and management instruments (SANCHEZ, 2008).

Because it is discussed in various areas of knowledge, environmental management has countless definitions, concepts and ways of being treated, but they all converge towards the same goal of corporate commitment to environmental issues. Among the most diverse definitions, the following stand out for the purposes of this study:

> Environmental management involves planning, organising and guiding the company to achieve specific [environmental] goals, in an analogy, for example, with quality management. A relevant aspect of environmental management is that its introduction requires decisions at the highest levels of management and therefore sends a clear message to the organisation that it is a corporate commitment. Environmental management can also become an important tool for organisations in their relations with consumers, the general public, insurance companies, government agencies, etc. (NILSSON, 1998, p. 134).

In this way, environmental management can be understood as the administration of environmental resources, with the main aim of conserving them and ensuring that future generations can enjoy an environment that is compatible with their needs, as can be found today.

Environmental management can also be conceptualised differently depending on the nature of the organisation, whether public or private. Public environmental management is a process of mediating interests and conflicts between social actors acting on the physical, natural and built environments. This mediation process continually defines and redefines the way in which the different social actors, through their practices, alter the quality of the environment and also how the costs and benefits resulting from the actions of these agents are distributed in society (IBAMA, 1995).

Public policies geared towards environmental management, in turn, should aim not only to manage resources to protect the natural environment, but also to serve as a guide in resolving social conflicts involving environmental issues, with a view to social well-being and the conservation of resources for future generations (FLORIANO, 2007).

Private environmental management, in turn, is an integral part of an organisation's overall management system, consisting of an administrative, dynamic and interactive process of resources, with the aim of balancing environmental protection and pollution prevention with socio-economic needs, as well as seeking to help protect human health, through the formulation of policies that can take into account the issue of information regarding significant environmental impacts, in accordance with society's own interest in the issue of environmental protection (FLORIANO, 2007).

In view of this, it can be said that the management of environmental resources must be supported by a strategic vision of long-term development, constituting the centre where the objectives associated with development and those aimed at conserving nature or preserving environmental quality meet and meet again.

However, this development needs to be sustainable. Sustainable development means development that can meet the needs of the present without compromising the ability of future generations to meet their own needs, since population growth is becoming constant and unlimited, while natural resources are limited.

Reflecting from the perspective that sustainable development suggests a permanent legacy from one generation to the next, so that all are able to provide for their needs, sustainability is seen as the conservation and maintenance of the natural resources available for human needs, requiring scientific and technological advances that allow for the permanent expansion of the capacity to utilise, recover and conserve these resources, as well as new concepts of human needs, in order to alleviate society's pressures on them (BARBIERI, 2011).

2.4 ENVIRONMENTAL IMPACTS

In the light of the information presented so far, it is important to discuss the issue of environmental impacts generated by man's actions in nature. These impacts are most often associated with damage to nature, such as the mortality of wildlife in the face of an oil spill at sea or in a river, when the image of birds completely blackened by the layer of oil covering them shocks public opinion (SÁNCHEZ, 2008).

Although this is man's action against nature, it is known that natural resources are finite in any case, for a relationship of infinite or constant needs for human survival, which only seeks to utilise them in a more balanced and consistent way. On this subject, Loureiro (2000) states that:

> The causes of environmental degradation and the crisis in the society-nature relationship do not emerge solely from circumstantial factors or humanity's perverse instinct, and the consequences of such degradation are not just the result of the misuse of natural resources, but rather a set of interconnected variables derived from the categories: capitalism/modernity/industrialism/urbanisation/technology.

In this way, the discussion goes beyond man's action in utilising the natural resources available around him and interacts with the social and production relationship. The generation of wealth through the practice of consumerism and buying and selling relationships encourages an increase in the uncontrolled exploitation of natural resources and, consequently, the degradation of the environment with the waste from this type of consumption.

The concept of environmental impact can be defined as "any alteration to the environment in one or more of its components caused by human action" (SANCHEZ, 2008). In this concept, it is important to note that the author relates environmental impact to human action, since the other species of living beings on the planet take from nature only what is essential for their consumption, do not stockpile for profit or waste the resources they use, thus damaging the natural cycle of fauna and flora or polluting their own habitat.

2.4.1 Environmental impacts in the ceramics industry

Environmental impacts cannot be categorised according to the nature of human activities, but rather their actual action on nature. In this sense, environmental impact can be defined as any type of change in the environment that occurs as a result of activities carried out by man or by an organisation, and its understanding as environmental impact will always be related to the generation of undesirable events that harm and compromise the environment (MOURA, 2004).

In addition to the environmental impact, dust emissions caused by the flow of vehicles at the site can occur throughout this process, generating visual impact, noise and vibrations, as well as the degradation of the working environment itself. According to Bitar (1997), with the presence of this type of degradation, native vegetation and local fauna begin to be destroyed and the quality and flow regime of the water system begins to change. According to this author, environmental degradation occurs when there is a loss of adaptation to physical, chemical and biological characteristics and socio-economic development becomes unfeasible.

According to data from Nascimento and Brito (2007), based on a study of the impacts generated by a ceramics industry in the Seridó region of the state of Rio Grande do Norte, it was found that the shrubs, considered to be one of the environmental factors analysed, were left unprotected after the trees were felled, and that these shrubs were left to serve as pastures for small herds of goats, which are usually placed to graze on the land exploited by the ceramics industry. When we look at the issue of birds, we can also see that they have been harmed by the felling of the trees, resulting in losses such as the loss of food sources and hiding places for these birds against their predators (NASCIMENTO; BRITO, 2007).

CHAPTER 3

MATERIALS AND METHODS

3.1 STUDY AREA

The municipality of Barcelona has 51 businesses, including 11 individual entrepreneurs, 12 linked to the public sector and 23 unincorporated businesses. Of these, 28 have their activities linked to commerce, 13 to structure, 2 to the industrial sector and 8 work with the provision of services. Retail is the most represented type of enterprise in the city, with 26 (PORTAL DO DESENVOLVIMENTO, 2015).

As it is a simple town, with a modest population of less than 5,000 inhabitants and a land area of 152,626 km^2, far from large urban centres, its way of life and socio-cultural development also influence the analysis and control of environmental impacts that may occur in that municipality, especially when there is any industry or commerce that can be translated as an opportunity for growth, employment and income for the local population (IBGE, 2010).

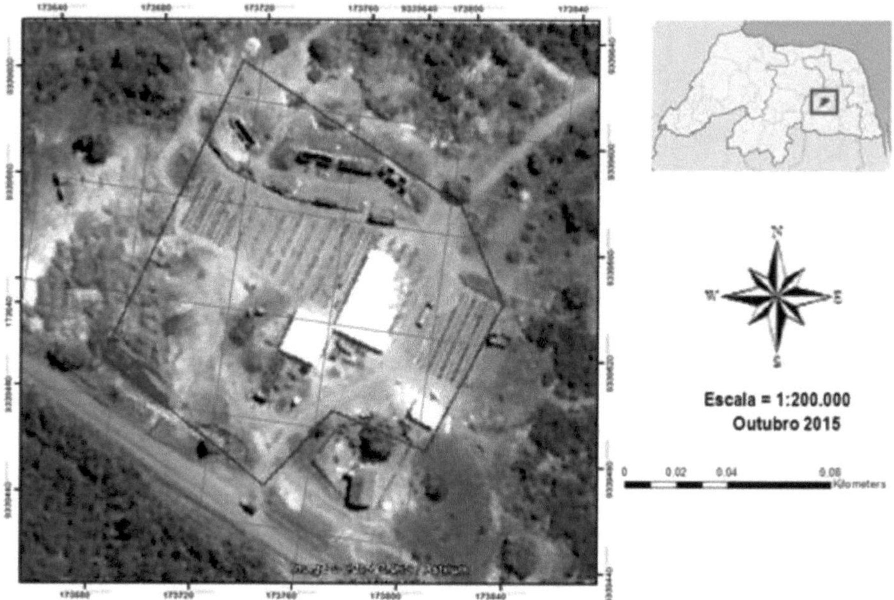

Figure 1 - Location of Barcelona Ceramics Ltd.
Source: Adapted from Google Earth (2015).

Within the scenario of different local activities and enterprises, there is the company Cerâmica Barcelona, located in a rural area approximately 5 kilometres from the town of Barcelona, in the state of Rio Grande do Norte, along the RN-203 motorway.

The company employs around 35 people and produces 500,000 pieces a month, with bricks being its only product. Slabs and roof tiles are also sold, but they are purchased from producers in the Seridó region of the state and are run by two brothers, i.e. it is a family business. The organisation also has a simple physical administrative structure, with just one office located in the production sector.

Figure 2 - Main sectors of Barcelona Ceramics Ltd.

Source: Adapted from Google Earth (2015).

They work 44 hours a week, from Monday to Friday, split into two 9-hour shifts so as not to work on Saturdays. Overtime is rare, but when it does occur it is paid in accordance with labour legislation.

The company doesn't have precise figures on the education level of its workers, but says that the majority of its employees are literate, ranging from primary school graduates to those who have completed high school.

In the field of workers' health, the company has a Medical Control and Occupational Health Programme (PCMSO), which aims to prevent, track and diagnose work-related health problems at an early stage. The company has personal protective equipment (PPE), which helps prevent accidents that may occur in the product's production processes.

On the technical side, the company doesn't have a ceramics technician, a professional who could bring benefits such as improving production rates and product quality, as well as optimising each production process.

The clay, an essential raw material in the production of bricks, is purchased from various suppliers in the Potengi region and is transported by the company's own vehicles. The firewood, used as fuel to burn the product, comes from algaroba, hazelnut, cashew and jurema trees, which are sourced from the local region and transported by the company's own vehicles. Of the suppliers, only two have a Forestry Origin Document (DOF), but they all supply the necessary firewood.

Figure 3 - Raw material used as fuel during the burning of the final product

Source: Data collected by the authors (2015).

The water used in the company's production comes from a tube well located on the River Potengi, which does not have a licence to use water resources. For human consumption, it comes from the Monsenhor Expedito Pipeline. The energy source used by the company is the public network, Cosern, which has an average consumption of 14,000 kW, and there is no alternative source.

The clandestine use of water can also be seen as an activity that could have an environmental impact, since there is no proper study or control of this type of activity. You would therefore need authorisation to drill the well and a study on the issue of the source and the appropriate flow rate.

During rainy periods, the company has a stock of raw materials so that production doesn't come to a halt, damaging production. It also has four kilns with a capacity of 30,000 bricks, which require 20 cubic metres of firewood to burn. Of the firewood suppliers, only two are authorised by the environmental control bodies, through a document called the Forestry Origin Document (DOF). There are around ten suppliers in total to supply the firewood needed for the burning process, which means that more than half of this process is also carried out illegally and clandestinely.

In terms of product quality, it is divided into 40 per cent first class, 40 per cent second class, 18 per cent third class and 2 per cent non-conforming material, which is used in a variety of activities, from building foundations to landfills, among others. The product meets the local needs of the municipality, the region and the Metropolitan Region of Natal. Complying with the requirements of ABNT and INMETRO legislation, the product comes with the CNPJ, name and date of manufacture stamped on the body of the product.

Before commenting on the environmental impacts caused by the ceramics industry, it is important to note how the production process is characterised in the manufacture of red ceramics, according to the flowchart shown in Figure 4. The main stages of the process are: extraction, disintegration, mixing, rolling, extrusion, cutting, drying, firing, inspection, storage and dispatch.

Next, it will be shown how the entire process from raw material extraction to final product storage is carried out at Indústria Cerâmica de Barcelona Ltda, so that this entire process can be understood and visualised, according to the surveys carried out at the company being researched.

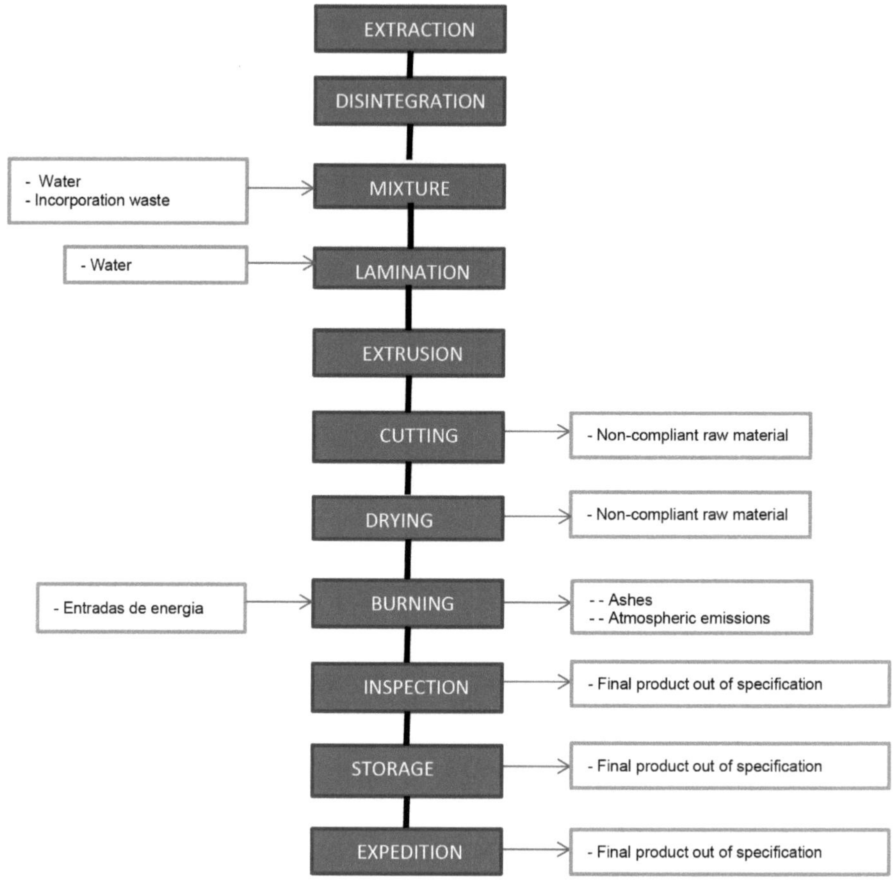

Figure 4 - Production process in the manufacture of red ceramics

Source: Adapted from the Brazilian Ceramics Association (2013).

The clay extraction process takes place in the open air, preferably in the months with the least rainfall, and can be carried out manually or mechanised with the help of excavators, wheel loaders, bulldozers with blades, and other equipment (see Figure 5). In several places where the raw material is extracted, the layer of sand that lies under the clay layer is also used. Even though it is not necessary for the ceramics industry, the sand extracted has a high economic value because it supplies the construction industry (FIEMG, 2013).

Figure 5 - Clay extraction at Barcelona Ceramics Ltd.

Source: Data collected by the authors (2015).

The stock of raw material can be the storage of all the raw material extracted from the quarries or the stock of clay that has already passed through the disintegrator. In the first case, the stock is in the pottery yard, out in the open, as shown in Figure 6. In the second case, the stock is in an appropriate, covered place, so as to obtain higher quality production, where it is disintegrated for a period of approximately one month.

The raw materials must be weighed and checked for accounting and production records before they enter the stock sector, in order to control the yield of the production process and, when necessary, to pay suppliers if these raw materials are purchased. This control will also serve to verify and control the quality of the raw material being used to manufacture the ceramic products (SEBRAE, 1999).

Figure 6 - Stock of raw materials at Barcelona Ceramics Ltd.

Source: Data collected by the authors (2015).

Mixing, this stage of the process takes place after storing the raw material collected to speed up the decomposition process of the organic matter and soluble salts still present in the sample and after a period of rest, the ceramic mass is prepared by mixing the various types of clay, water and waste (see Figure 7), where applicable, seeking to produce a good ceramic mass with the main objective of obtaining a product that has high quality, reducing losses and improving the environmental performance of the process (FIEMG, 2013).

The ceramic mixture is prepared using a mixer in an average ratio of one part water to three parts clay. The ceramic mixture should be thin, reducing shrinkage and preventing the pieces from deforming and cracks appearing (SEBRAE, 1999).

The main objective of clay and dough preparation is to obtain a homogeneous mixture, with constant characteristics and a suitable unit for the particular forming method used. In this way, the quality of the product to be acquired depends directly on the control that must be carried out at the various stages of preparation, through continuous control, so that quality can be ensured and maintained at the desired level (BASTOS, 2003).

Figura 7 - Wetting for mixing the raw material at Barcelona Ceramics Ltd.

Source: Data collected by the authors (2015).

Figura 8 - Mixing raw materials at Barcelona Ltda ceramics plant

Source: Data collected by the authors (2015).

Disintegration consists of reducing the clods to dimensions that allow them to be fed and operated by the plant's mill, increasing the homogeneity of the mass (Figure 9). This stage of the process is also known as crushing and grinding.

The most common types of equipment used for this purpose are jaw mills and cage mills. The suitability of the type of equipment depends on the initial size of the clay blocks or clods and the desired degree of disintegration (MINEROPAR, 2007).

Figura 9 - Disintegration of raw materials at ceramics company Barcelona Ltda

Source: Data collected by the authors (2015).

Rolling consists of directing the clay particles, which is done by periodically adjusting the rolling equipment (Figure 10). The more closed the rolling mill is, the better the particles will be directed, since the quality of the rolling determines the quality of the finish of the products, avoids losses and contributes to reducing energy consumption losses during firing, since the granulometry of the material decreases (FIEMG, 2013).

The laminators have the function of reducing and homogenising the thickness of the piece of clay material that enters the maromba (to be explained later), making it easier to expel the air through the maromba's vacuum. The mixed and laminated material is then transported to the extruder or maromba, where the clay material enters in its raw form and the product comes out in its final form, or in the form desired so that it can be sent to the presses, as in the case of roof tiles (SILVA, 2012).

Figura 10 - Lamination of raw materials at Barcelona ceramics

Source: Data collected by the authors (2015).

Extrusion is the process in which the plastic mass is placed in an extruder, also known as a maromba, where the mass is compacted and forced by a piston or helical shaft, through a nozzle with a certain shape, obtaining an extruded column, with a cross section, with the desired shape and dimensions (Figure 11). This column is then cut and parts such as hollow bricks, blocks and other regular shaped products are obtained. Extrusion is often seen as the intermediate stage of the forming process, followed by the cutting of the extruded column used in the manufacture of roof tiles or turning, as well as the manufacture of electrical insulators, cups and plates, among other products (ABCERAN, 2011).

This is a plastic forming process limited to the manufacture of objects with a constant cross-section. At this stage of the process, the main factors affecting the yield and quality of the final products are: the composition and preparation of the ceramic masses; the plasticity of the masses; the extrusion moulds (nozzles); and the type of extruder and the quality of the vacuum. Thus, although the costs of preparing and shaping the raw materials in the ceramics industry are relatively low in relation to the total costs of making the product, if these operations are not carried out correctly, they can cause a significant increase in the costs of drying and firing, which are the most expensive stages of the whole process (WIECK; DUAILIBI FILHO, 2014).

Figura 11 - Extrusion of raw materials at ceramics company Barcelona Ltda

Source: Data collected by the authors (2015).

The raw material is cut at the exit of the extruder or maromba. The pieces are cut in equipment called cutters (see Figure 12), which can be automatic or manual, in their final form, as in the case of bricks, or in the form of sticks, known as pastões, with a square, cylindrical, concave or other desired shape, to be subsequently pressed into their final form, as in the case of roof tiles (SILVA, 2012).

The cutting table is located right at the extruder's flow outlet and is made up of wires attached to a metal bracket. The cut is synchronised and precise, and is distinguished by its horizontal cutting system, which reduces burrs and deformations in the clay elements. After this process, the products are transported by specific hand trolleys to the product drying area, which is now semi-finished (MACIEL, FREITAS, 2013).

Figura 12 - Cutting raw materials at Barcelona Ceramics Ltd

Source: Data collected by the authors (2015).

The final product can be dried directly on the floor or on shelves, depending on how the company usually carries out this process and the climate (Figures 13 and 14). In some more developed potteries, there are kilns suitable for this purpose and the pieces are transported using wagons (SILVA, 2012).

Figure 13 - Drying the final product at the Barcelona Ltda ceramics plant

Source: Data collected by the authors (2015).

Figura 14 - Final product drying yard at Barcelona Ceramics Ltd.

Source: Data collected by the authors (2015).

The firing of the final product takes place in a kiln, which is seen as the central element of the pottery (Figure 15), where the ceramic material is fired at a temperature varying between 750°C and 1,000°C, passing through the stages of preheating (with a temperature of up to 200°C), low fire (with a temperature between 200°C and 600°C) and high fire

(with a temperature above 600°C up to the maximum limit required for burning), temperature sustaining and cooling (SILVA, 2012). Vegetable biomass is often used as a source of thermal energy (Figure 16).

This stage is also known as synthesising, where the products acquire their final properties and the pieces, after drying, are subjected to heat treatment and high temperatures. During the firing process, a series of transformations take place depending on the components of the mass, such as loss of mass, development of new crystalline phases, formation of a glassy phase and welding of the grains (SINDICERMF, 2015).

After the firing and cooling process, the unmoulded ceramic blocks are ready to be sold. This stage is of great importance in this production process and requires special attention because, without proper care, it can lead to the total loss of a batch. In order to avoid this loss, there is a need to control the firing time, temperature and internal ventilation of the kiln (GALASSI; TAVARES, 2013).

Figura 15 - Kiln where the final product is fired at the Barcelona Ltda ceramics plant

Source: Data collected by the authors (2015).

Figura 16 - Biomass used in the firing of the final product at the Barcelona Ltda ceramics plant

Source: Data collected by the authors (2015).

Inspection of the final product is carried out as soon as the finished product leaves the kiln, by the deforgers, where materials with defects are separated (Figure 17). These rejected products can be reused and reprocessed, depending on the type and defect found, or used to be sold to the construction industry (GARCIA; RIBEIRO; OLIVA, 2009).

At Cerâmica Barcelona, we can see that the average percentage of products rejected due to defects that are reused in the production process itself is just 2% (see Figure 18). Of this percentage, 2% is reused in the production process and the rest is sent to the construction industry to be used in landfills, road maintenance, among others.

Figura 17 - Removal of the final product for further inspection at Barcelona Ceramics Ltd.

Source: Data collected by the authors (2015).

Figura 18 - Waste from the dispatch of the final product at ceramics company Barcelona Ltda

Source: Data collected by the authors (2015).

The final product is dispatched by lorry, by road, using the company's own equipment or drivers. In some regions of Brazil, ferries are also used when the location is close to rivers. The material

stored in a covered or uncovered area, depending on the climatic conditions of each region, they remain in the storage sector until they are authorised to be removed by dispatch (GARCIA; RIBEIRO; OLIVA, 2009).

Storage of the finished product may come before or instead of inspection in certain companies, or it may come before dispatch. However, regardless of the order of the stages, all finished products that have not been dispatched are stored in the open (Figure 19), on a platform at the side of the kiln, with no identification of quantity or quality. In most cases, while the lorry is being loaded, the driver goes to the office to request the invoice for the products to be transported. After the lorry is loaded, the product to be transported is checked by the person in charge of the dispatch department and then released for delivery (MACIEL; FREITAS, 2013).

In the case of Cerâmica Barcelona, the separation of first, second and third quality products is carried out at the time of inspection, which facilitates the process of storage and dispatch of the finished products, since they will be marketed according to their quality.

Figura 19 - Finished product storage yard at Barcelona Ceramics Ltd.

Source: Data collected by the authors (2015).

3.2 METHODOLOGY

3.2.1 Sources used

This research study used primary sources to collect information through an *on-site* visit using interviews with the managers of the company Cerâmica Barcelona Ltda., as well as photographic records of the same, and secondary sources through a survey of the theoretical framework previously explained and other additional information on the research environment and the subject studied, considered relevant for its preparation.

A checklist was also used to assess the relevant and existing items regarding exploration activities and environmental management activities (Annex).

3.2.2 Type of research

As for the technique, a case study was carried out involving the company Cerâmica Barcelona Ltda., based in the municipality of Barcelona-RN, where the aim was to find out and evaluate the entire process of manufacturing the red ceramics produced by the company and the main stages of the process that could cause damage to the environment, generating a negative environmental impact.

The analysis was not concerned with carrying out a detailed study of all the strengths and weaknesses, but with identifying possible threats to the environment that could have a negative impact on the population of the town where the company is based, as well as on neighbouring towns, since much of the raw material used to cook the finished products, wood, is extracted in other towns.

3.2.3 Data processing

The data collected was processed using a qualitative analysis, through an exploratory study of the views of the managers interviewed in relation to the issues in this study. After feeding in the matrix used, graphs will be drawn up on the environmental impacts observed, using a Microsoft Excel 2010 spreadsheet, and after tabulating this data, a qualitative analysis will be carried out on the impacts observed *in loco*. It is worth emphasising that these observations cannot be considered exact data, since it depends on observation and the researcher's level of knowledge and understanding, and these data tend to be somewhat subjective, so that if other research is carried out in this area on the same subject, different results may be found.

According to Gil (2010), qualitative research takes into account the relationship between the objective of the study and the subjectivity of the researcher, an assessment that cannot be translated into numbers and the data is analysed using statistical instruments to describe the reality found.

The Leopold Method will be used as a method for assessing environmental impacts, so that it is possible to identify, collect and organise the data collected in the research environment related to the topic, allowing the data to be presented in formats that enable it to be better interpreted. The Leopold matrix makes it possible to point out all the possible interactions between actions and factors, and then establish the magnitude and importance of each impact on a scale ranging from 1 to 3, identifying whether it is positive or negative. The magnitude can be assessed objectively or empirically, as it refers to the degree of change caused by the action on the environmental factor, while the importance score is subjective or normative, as it involves assigning relative weight to the factor affected by the project (COSTA E CHAVES, 2005).

CHAPTER 4

RESULTS AND DISCUSSION

During the month of September, several visits were made to the research organisation, with the main aim of documenting all the stages of the production process through photographic records and, at the same time, collecting data for the preparation of the Leopold Matrix, obtaining the following data and information, as can be seen in Tables 1 and 2.

In Figure 20, the impacts presented are all specified in the Matrix presented in the annex, with their respective evaluations: positive, negative and neutral. The Leopold Matrix was used to assess the following natural and human elements presented in Table 1:

Table 1 - Leopold Matrix

<table>
<tr><td colspan="21">Natural and Human Elements</td></tr>
<tr><td colspan="7">Physical and Chemical Characteristics</td><td colspan="5">Biological conditions</td><td colspan="7">Cultural factors</td><td colspan="2"></td></tr>
<tr><td colspan="2">1 wrong</td><td colspan="2">Water</td><td>Atmoslera</td><td colspan="2">Processes</td><td colspan="2">Flower"</td><td colspan="3">Fauna</td><td>Use of tefrit</td><td colspan="4">luieiesses Hionanõs and Aesthetics</td><td colspan="2">Level C II 1 lurai</td><td colspan="2">Ecological Relationships</td></tr>
<tr><td>Soils</td><td>Mineral recovery</td><td>Superficial</td><td>Quality</td><td>Quality (gases, particulates)</td><td>Erosion</td><td>Compacting</td><td>Trees</td><td>Shrubs</td><td>Birds</td><td>Land animals including reptiles</td><td>Insects</td><td>Open and wild spaces</td><td>Panoramic views and landscapes</td><td>Quality of open spaces</td><td>Landscape design</td><td>Nature (forest quality)</td><td>Cultural patterns ()</td><td>Jobs</td><td>Food chains</td><td>Disease vectors (insects)</td></tr>
</table>

Source: Adapted by the authors (2015).

Table 2 - Impact Valuation

Impact valorisation			
Character (Ca)	Positive (1)	Neutral (0)	Negative (-1)
Importance (I)	High (3)	Average (2)	Low (1)
Coverage (C)	Regional (3)	Location (2)	Punctual (1)
Duration (D)	Permanent (3)	Average (2)	Short (1)
Reversibility (R)	Irreversible (3)	Partial (2)	Reversible (1)
Total Impact (IT) = IT = Ca. (I + Co + D + R)			

Source: Adapted by the researcher (2015).

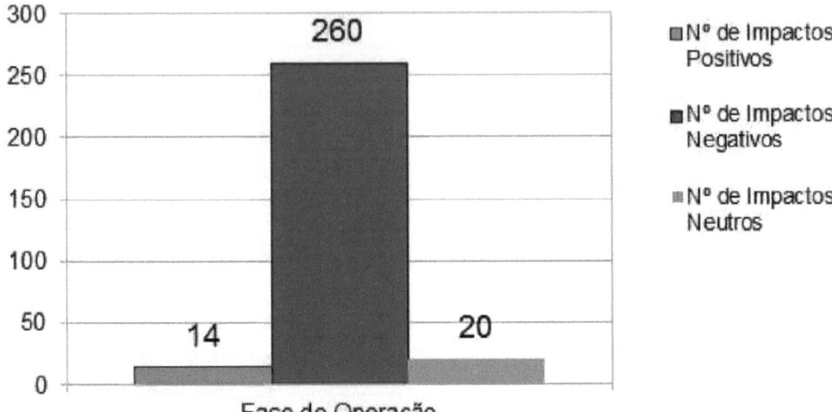

Figure 20 - Number of environmental impacts

Source: Prepared by the authors (2015).

According to the data shown in Figure 20, it can be seen that throughout the red ceramics production process there are a fairly high number of negative impacts, compared to positive and neutral impacts.

According to the data presented in the matrix (appendix), the only positive impacts are employability and cultural change. Almost all the other impacts are negative, at all stages of the process, with only noise, vibration and changes in drainage into the water and atmosphere being considered neutral. This data reveals the need to take a closer look at the issue of maintaining natural resources and their sustainable use, which are being used in a very disproportionate way and which could cause major damage to society in terms of quality of life in the medium term. These impacts will be discussed in the next graphs according to the table provided in the annex and their respective data evaluated in Figures 21, 22 and 23.

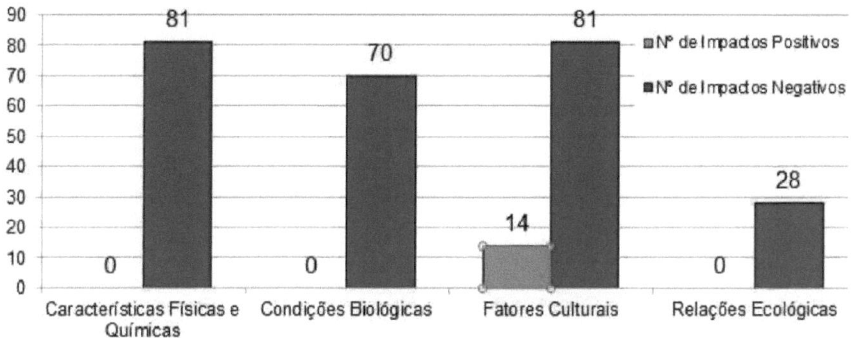

Figure 21 - Number of impacts in the operation phase per impacted medium

Source: Prepared by the authors (2015).

According to the data presented in Figure 21, it can be seen that all the items related to the environment appear negative, with no action being taken to minimise this ecological imbalance, such as the lack of reforestation of deforested areas, the reduction of wood combustion gases during burning, among others. The only positive impact is the issue of employability, which means that the more the company grows, the more people will find work.

However, this is a factor that generates more negatives than positives, since these people will not have a quality of life, deforestation will be carried out in ever greater proportions, causing animals to migrate, more toxic gases will be emitted into the atmosphere, in addition to the use of the raw material, clay, in ever greater quantities, resulting in an environment that is favourable to soil erosion, as discussed above in section 2.4.1.

During the operation phase or production process of red ceramics (Figure 22), it can be seen that there are many more negative impacts than positive ones for the environment and society in general. Generating jobs and income without quality of life cannot be seen as a positive point, since a sick society represents a high cost for the health sector of the community where the company is located or where the wood and raw materials for the production process are extracted.

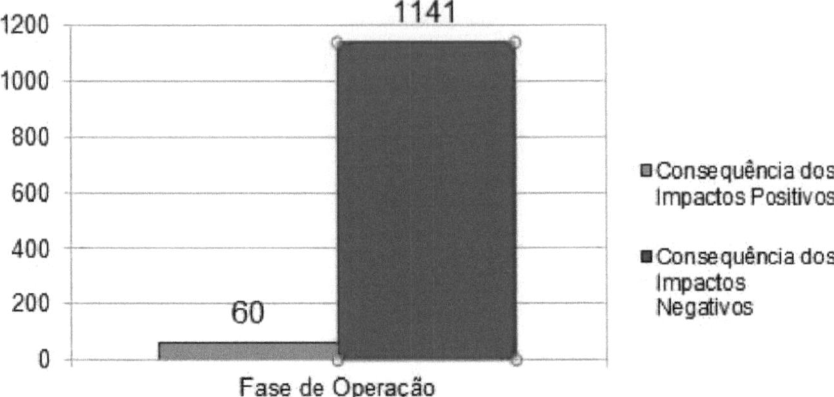

Figure 22 - The consequences of environmental impacts in the operational phase

Source: Prepared by the authors (2015).

In this context, there needs to be monitoring by the bodies responsible for preserving the environment, specific laws and, if necessary, the use of Conduct Adjustment Agreements (TACs), so that these companies can adapt to environmental requirements in such a way that they don't damage people's quality of life, fauna and flora. It's worth pointing out that TACs are only thought of or used when a company or group of companies is having a major negative impact on the environment, be it the soil, the atmosphere, the fauna, among others. The bodies mentioned above, according to their remit, are: IBAMA, IDEMA, the Municipal Department of the Environment, the Environmental Police, among others.According to the data in Figure 23, it can be seen that the impacts throughout the operating phase have a very negative and worrying characteristic, which is only observed in studies such as this one.

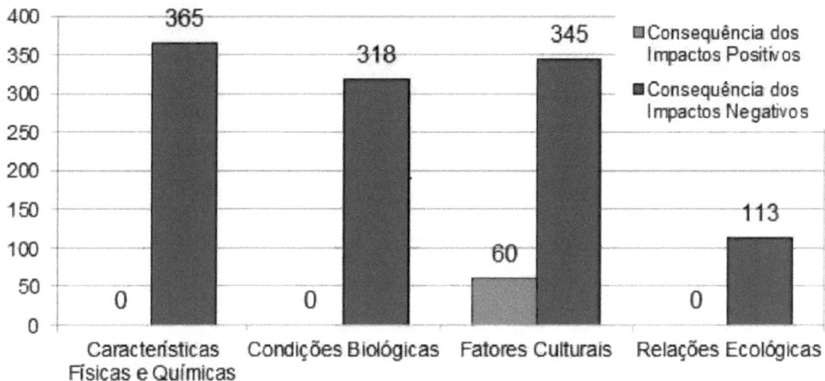

Figure 23 - Consequence of impacts in the operation phase

Source: Prepared by the authors (2015).

The physical and chemical conditions become noticeable during the burning of the wood in the drying process of the final product and in the areas where the clay is extracted for production. The carbon dioxide inhaled by workers, inhabitants and the fauna present in the environment becomes a biological danger to the lives of all beings present in this context. As the environment changes, so does the habitat itself and so does the migration of birds and animals. Even though ceramic production generates jobs and income and can even improve the conditions of some families, the damage caused by cultural factors is far greater than the positive aspects. In short, the ecological relationship is directly linked to food chains and disease vectors, as well as the presence of insects, which contributes to an ecological imbalance in this research environment.

For those who work at the company or live near it, these impacts are not noticeable in a short space of time, except in the absence of certain types of animals that inhabit the region, the presence of soil susceptible to erosion or the deforested areas that are gradually appearing. The most dangerous and least noticeable impact is the emission of toxic gases into the atmosphere during the burning of wood to cook the final product, as these gases rise into the atmosphere, contributing to the greenhouse effect and the reduction of the ozone layer, and can often precipitate in the rain, causing great harm to health and the soil.

It can be seen that the most negative impact is related to the extraction of clay, as can be seen in the matrix provided in the annex, since it could favour erosion and make it difficult to recover the area used. The same applies to the wood used to cook the final product, since there is no reforestation process or deforestation control.

At the company we surveyed, only two wood suppliers have the Forestry Origin Document (DOF), which allows them to extract and sell the wood to be used in the pottery. However, it is clear that other suppliers who do not have this type of authorisation also sell.

CHAPTER 5

CONCLUSION

In the light of the data and information analysed, it can be seen that the process of manufacturing products using red ceramics causes a great deal of damage to the environment, which should be observed and discussed more frequently by researchers, the supervisory bodies responsible and society in general. It is an activity that is very useful to society and almost irreplaceable in its general use in civil construction. It is also an industrial sector that is in an expansion phase, given the demographic growth of many cities across the country.

However, at the very least, there needs to be more strategic planning about the negative consequences for the environment, a review of the way wood is acquired from suppliers and marketing only with suppliers who have a DOF, as it is easier to control their actions. The way clay is extracted should be rethought, so as to reduce the possibility of future erosion or regeneration of the environment in which the raw material is being extracted.

The emission of polluting gases into the atmosphere, such as carbon dioxide, for example, is an action that can be managed according to the cooking process of the final products. If possible, work can be done to replace the whitewash ovens with others that emit fewer pollutants and use other energy sources, avoiding deforestation and achieving the desired temperature.

The positive aspect is the generation of jobs and income for the town and region where the company is located. However, the number of people working and the company's level of production must be monitored, since the more it produces, the more negative consequences there will be for the environment and the quality of life of people in the region. This can even lead to serious health problems related to the respiratory system through the inhalation of toxic gases.

The extraction of wood also contributes to the migration of the animals that inhabit the region, since the native vegetation is being removed and there is no reforestation process on the part of the wood suppliers. Even if this process took place, as it involves specific types of vegetation for burning, the soil would gradually lose its primary composition and would no longer be suitable for this process, due to the lack of nutrients needed for the new trees to grow. What's more, this is a process that needs to be analysed in terms of the time needed for each type of tree felled to reach the adult stage and be harvested again.

Finally, there is the issue of noise that is observed during the operation of machines and equipment used during the production process. However, this phenomenon cannot be considered very serious, since if the production process ceases, the noise will cease, and it will only affect those inside the company.

REFERENCES

ABCERAM. Brazilian Ceramics Association. **Technical information**: manufacturing processes. 2011. Available at: < http://www.abceram.org.br/site/?area=4&submenu=50 > Accessed on: 29 Sep. 2015.

. Brazilian Ceramic Association. **Ceramics in Brazil**: Technical Standards. Available at:< http://www.abceram.org.br/site/index. php?area=2&submenu=18>. Accessed on: 17 August 2015.

AFANCER. **History of Ceramics**. 2015. Available at: <http: // www.anfacer.org.br/site/default.aspx?idConteudo= 157&n=Hist%C3 %B3ria- da-Cer%C3%A2mica> Accessed on: 14 July 2015.

BARBIERI, José Carlos. **Development and the Environment**. Agenda 21's strategies for change. 12.ed. Petrópolis: Vozes, 2011.

BASTOS, Frederico Assis. **Evaluation of the ceramic tile and block manufacturing process with a view to product certification**. Master's dissertation. Postgraduate Programme in Civil Engineering. Federal University of Santa Catarina. Floranópolis, 2003. 164 fl.

BITAR, O.Y. **Evaluation of the recovery of areas degraded by mining in the RMSP**. Thesis (Doctorate), Department of Mining Engineering, Polytechnic School, University of São Paulo. São Paulo, 1997.

BYLAARDT, Marina Paulino et all. **The origin of ceramics**. Experimental Project Handicrafts. UFMG. Available at:

<http: //www.eba.ufmg.br/aluno s/kurtnavigator/arteartesanato/origem.html> Accessed on: 5 Aug. 2015.

BUSTAMANTE, Gladstone Motta; BRESSIANI, José Carlos. **The Brazilian ceramics industry**. Cerâmica Industrial, vol. 5, May/June 2000. Available at:< http://ceramicaindustrial.org.br/pdf/v05n03/v5n3_5.pdf> Accessed on: 8 Sep. 2015.

COSTA M; CHAVES, P. **Uso das Técnicas de Avaliação de Impacto Ambiental em Estudos Realizados no Ceará**. Paper presented at NP 09 - Scientific and Environmental Communication, of the V Meeting of Inercom Research Centres, XXVIII Brazilian Congress of Communication Science, Rio de Janeiro. 2005.

FIEMG. Federation of Industries of the State of Minas Gerais. **Environmental technical guide for the red ceramics industry**. Belo Horizonte, 2013. Available at: <http: //www.feam.br/image s/storie s/producao_sustentavel/GUIAS_TECNICOS_AMBIENTAIS/guia_ceramica.pdf> Accessed on: 21 Aug. 2015.

FIERN/SENAI. **Industrial profile of red ceramics in Rio Grande do Norte**. Natal/RN: SENAI, 2001. 75p.

FLORIANO, Eduardo Pagel. **Policies in Environmental Management**. 3.ed. Santa Maria: UFSM-DCF, 2007

GALASSI, Cristiane; TAVARES, Célia Regina Granhen. **Ceramic block production process**. Maringaense Production Engineering Symposium. Management and innovation in the production chain. 2013. Available at: <http://webcache.googleusercontent.com/search?q=cache:KQ_KHtDQH5UJ:www.dep.uem.br/simepro/anais/index.php/simepro/6_simepro/paper/download/9/10+&cd=1&hl=en-BR&ct=clnk&gl=br> Accessed on: 4 Oct. 2015.

GARCIA, Fabrício de Almeida; RIBEIRO, Marcela Bellucci; OLIVA, Rafael K. Bonizio. **Analysing the profile of the red ceramics industry using the social carbon methodology**. 2009. Available at: http://www. socialcarbon.org/wp-content/themes/socialcarbon/docs/KUPPER CERAMIC SOCIALCARBON.pdf Accessed on: 6 Oct. 2015.

IBGE. **Rio Grande do Norte**: Barcelona. 2010. Available at: < http://www.cidades.ibge.gov.br/xtras/perfil.php?lang=&codmun=240150&search=ri o-grande-do-norte|barcelona|infograficos:-informacoes-completas> Accessed on: 20 Jun. 2015.

KEIMERICH, Pedro Daniel da Cunha et all. Assessment of environmental impacts in the implementation and operation of pottery. Engenharia Ambiental Espírito Santo do Pinhal, v.8, n.1, p. 134-150, jan./mar., 2011. Available at: <http://ferramentas.unipinhal.edu.br/engenhariaambiental/include/getdoc.php?id=143 3&article=502&mode=pdf.> Accessed on: 11 October 2015.

LEGGERINI, Maria Regina Costa. **Ceramic Materials**. 16 September 2010. Available at: <http: // www.feng . pucrs .br/professores/mregina/ARQUITETURA_-_Materiais_Tecnicas_e_Estruturas_I/estruturas_i_capitulo_II_materiais_ceramicos.p df> Accessed on: 25 October 2015.

LOUREIRO, Carlos Frederico Bernardo. Social theory and the environmental question: presuppositions for a critical praxis in environmental education. In: LOUREIRO, Carlos Frederico Bernardo et al. **Sociedade e meio ambiente**: a educação ambiental em debate. São Paulo: Cortez, 2000.

MINEROPAR. **The preparation of clays for the production of ceramic tiles and blocks**. Curitiba, 2007. Available at: <http: // www.mineropar . pr. gov. br/arquivo s/F ile/publicacoe s/Prepara_Argilas_Produc ao_Telhas_Blocos_Ceramicos_2007.pdf> Accessed on: 21 Aug. 2015.

MOURA, Luiz Antonio Abdalla de. **Quality and Environmental Management**. São Paulo: Editora Juarez de Oliveira, 2004.

NASCIMENTO, Waldécio Sávio dos Anjos do. **Evaluation of the environmental impacts generated by a typical ceramics industry in the Seridó/RN region**. Master's dissertation. Federal University of Rio Grande do Norte. Postgraduate Programme in Mechanical

Engineering. Natal, 2007. Available at: <repositorio.ufrn.br/jspui/bitstream/123456789/15735/1/WaldecioSAN.pdf> Accessed on: 11 October 2015.

NILSSON, W. R. Services instead of products: experiences from energy markets - examples from Sweden. In: MEYER-KRAHMER, F. (Ed.). **Innovation and sustainable development**: lessons for innovation policies. Heidelberg: Physica-Verlag, 1998.

POINT OF ART. **The art of ceramics**. 16 June 2011. Available at: <> http://pointdaarte.webnode.com.br/news/historia%20da%20cer%C3%A2mica/Accessed on: 6 Aug. 2015.

VIRTUAL PORTAL OF BARCELONA. **History**. Available at:< http://www.barcelona.educ.ufrn.br/historia.htm> Accessed on: 20 June 2015.

DEVELOPMENT PORTAL. **Barcelona**. Available at: <> http://www.portaldodesenvolvimento.org.br/upload/rn_pdfs/BARCELONA.pdfAccessed on: 21 Jun. 2015.

SANCHEZ, Luis Enrique. **Environmental impact assessment**: concepts and methods. São Paulo: Oficina de Textos, 2008.

SEBRAE. **Ceramic products**. Project profile series. Vitória. Dec. 1999. Available at: <http://201.2.114.147/bds/bds.nsf0D789DFB988A1A858325742F00556B0E/$File/P rodutos%20cer%C3%A2micos.pdf> Accessed on: 15 Sep. 2015.

. **Red Ceramics for Construction**: Tiles, Bricks and Pipes. SEBRAE/ESPM Market Study. Sept. 2008. Market Series. Available at: < http://201.2.114.147/bds/bds.nsf/C5B4284E12896289832574C1004E55DA/$File/N T00038DAA.pdf > Accessed on: 14 July 2015.

. **How to set up a ceramics industry**. 2014. Available at: <http://webcache.googleusercontent.com/search?q=cache:ZKbVHa66t28J:www.sebr ae .com. br/site s/Portal S ebrae/ideias/Como -montar- uma-ind%25C3 %25BAstria-de-

cer%25C3%25A2mica+&cd=2&hl=en-BR&ct=clnk&gl=br> Accessed on: 25 Oct. 2015.

SILVA, Isabelle Caroline Santos da. **Red Ceramics**. Federal Institute of Education, Science and Technology of Pernambuco. Academic Department of Infrastructure and Civil Construction. Recife, Oct. 2012. Available at: <> http://www.ebah.com.br/content/ABAAAfYZcAC/ceramica-vermelhaAccessed on: 2 Oct. 2015.

SILVA, Valdenildo Pedro. **Environmental impacts of the expansion of red ceramics in Carnaúba dos Dantas-RN**. Holos magazine, year 23, vol. 32007. Available at: <http://www.redeaplmineral. org.br/biblioteca/ceramica- vermelha/133-418-1-PB.pdf> Accessed on: 15 July 2015.

SINDICERMF. Red Ceramic Industry Union. **Ceramic Manufacturing Process**. Available at:< http://www.sindicermf.com.br/processo-fabricacao- ceramica.html> Accessed on: 4 October 2015.

APPENDIX

Matriz de Leopoldo - Environmental impacts in the ceramics industry Barcelona Ltda

		Elementos Naturais e Humanos																				
		Características Físicas e Quinicas							Condições Biológicas					Fatores Culturais							Relações Ecológicas	
		Terra		Água		Atmosfera	Processos		Flora		Fauna			Uso do território	Interesses Humanos e Estéticos				Nível Cultural			
		Solos	Recurso mineral	Superficial	Qualidade	Qualidade (gases, particulados)	Erosão	Compactação	Árvores	Arbustos	Aves	Animais terrestres, incluindo os répteis	Insetos	Espaços abertos e selvagens	Vistas e paisagens panorâmicas	Qualidade de espaços abertos	Desenho da paisagem	Natureza(qualidade da floresta)	Padrões Culturais (estilo de vida)	Empregos	Cadeias alimentares	Vetores de doenças (insetos)
Extração da matéria prima	Modificação de Habitat	-7	-7	-4	-4	-4	-4	-4	-7	-7	-7	-6	-6	-5	-5	-6	-5	-5	-4	7	-4	-4
	Alteração da Cobertura do Solo	-7	-7	-6	-5	-4	-6	-4	-7	-7	-6	-7	-7	-5	-5	-6	-7	-5	4	5	-5	-4
	Paisagens	-6	-6	-4	-4	-4	-5	-4	-6	-6	-5	-5	5	-4	-5	-4	-7	5	-4	0	-4	-4
	Escavações superficiais	-7	-7	-5	-4	-4	-4	-4	6	-4	-4	-4	-4	-4	-4	-4	-4	-4	-4	6	-4	-4
	Alteração da Drenagem	-4	-4	-4	-4	0	-4	-4	-4	-4	-4	-4	-4	-4	-4	-4	-4	-4	-4	4	-4	-4
	Ruidos ou Vibração	-4	-4	0	0	0	-4	-4	-4	-4	-4	-4	-4	-4	-4	-4	-4	-4	-4	-4	-4	-4
Desintegração	Ruídos ou Vibração	-4	-4	-4	-4	-4	-4	-4	-4	-4	-4	-4	-4	-4	-4	-4	-4	-4	-4	4	-4	-4
Laminação	Ruidos ou Vibração	-4	-4	0	0	-4	-4	-4	-4	-4	-4	-4	-4	-4	-4	-4	-4	-4	-4	4	-4	-4
Extrusão	Ruídos ou Vibração	-4	-4	0	0	0	0	-4	-4	-4	-4	-4	-4	-4	-4	-4	0	-4	-4	4	-4	-4
Corte	Ruídos ou Vibração	-4	-4	0	0	-4	0	-4	-4	-4	-4	-4	-4	-4	-4	-4	0	-4	-4	4	-4	-4
Mistura	Ruídos ou Vibração	-4	-4	0	0	0	-4	-4	-4	-4	-4	-4	-4	-4	-4	-4	1	-4	-4	4	-4	-4
Queima	Emissão de gases residuais	-8	-7	-4	-4	-6	-4	0	-5	-5	-4	-4	-4	-4	-4	-4	-4	-4	-4	4	-4	-4
	Ruídos ou Vibração	-4	-4	-4	-4	-7	-4	-4	-4	-4	-4	-4	-4	-4	-4	-4	-4	-4	-6	4	-4	-4
Estocagem	Armazenamento de produtos	-4	-4	-4	-4	-4	-4	-4	-4	-4	-4	-4	-4	-4	-4	-4	-4	-4	-4	5	-4	-4

Impact valuation table

Modificação de Habitat		Caráter (Ca)	-1	-1	-1	-1	-1	-1	-1	-1	-1	-1	-1	-1	-1	-1	-1	-1	-1	-1	1	-1	-1
		Importância (I)	2	2	1	1	1	1	1	2	2	2	1	1	1	1	1	1	1	1	2	1	1
		Cobertura (Co)	1	1	1	1	1	1	1	1	1	1	1	1	1	1	1	1	1	1	1	1	1
		Duração (D)	2	2	1	1	1	1	1	2	2	2	2	2	1	1	2	2	1	1	2	1	1
		Reversibilidade (R)	2	2	1	1	1	1	1	2	2	2	2	2	2	2	2	1	2	1	2	1	1
		Total	-7	-7	-4	-4	-4	-4	-4	-7	-7	-7	-6	-6	-5	-5	-6	-5	-5	-4	7	-4	-4
Alteração da Cobertura do Solo		Caráter (Ca)	-1	-1	-1	-1	-1	-1	-1	-1	-1	-1	-1	-1	-1	-1	-1	-1	-1	1	1	-1	-1
		Importância (I)	2	2	1	1	1	2	1	2	2	2	2	2	1	1	2	2	1	1	1	1	1
		Cobertura (Co)	1	1	1	1	1	1	1	1	1	1	1	1	1	1	1	1	1	1	2	1	1
		Duração (D)	2	2	2	1	1	1	1	2	2	1	2	2	1	1	1	2	1	1	1	1	1
		Reversibilidade (R)	2	2	2	2	1	2	1	2	2	2	2	2	2	2	2	2	2	1	1	2	1
		Total	-7	-7	-6	-5	-4	-6	-4	-7	-7	-6	-7	-7	-5	-5	-6	-7	-5	4	5	-5	-4
Paisagens		Caráter (Ca)	-1	-1	-1	-1	-1	-1	-1	-1	-1	-1	-1	-1	-1	-1	-1	-1	-1	-1	0	-1	-1
		Importância (I)	2	2	1	1	1	2	1	2	2	2	2	2	1	1	1	2	1	1	1	1	1
		Cobertura (Co)	1	1	1	1	1	1	1	1	1	1	1	1	1	1	1	1	1	1	2	1	1
		Duração (D)	2	2	1	1	1	1	1	2	2	1	1	1	1	2	1	2	2	1	1	1	1
		Reversibilidade (R)	1	1	1	1	1	1	1	1	1	1	1	1	1	1	1	2	1	1	1	1	1
		Total	-6	-6	-4	-4	-4	-5	-4	-6	-6	-5	-5	-5	-4	-5	-4	-7	-5	-4	0	-4	-4
Escavações superficiais		Caráter (Ca)	-1	-1	-1	-1	-1	-1	-1	-1	-1	-1	-1	-1	-1	-1	-1	-1	-1	-1	1	-1	-1
		Importância (I)	2	2	2	1	1	1	1	1	1	1	1	1	1	1	1	1	1	1	2	1	1
		Cobertura (Co)	1	1	1	1	1	1	1	1	1	1	1	1	1	1	1	1	1	1	1	1	1
		Duração (D)	2	2	1	1	1	1	1	2	1	1	1	1	1	1	1	1	1	1	2	1	1
		Reversibilidade (R)	2	2	1	1	1	1	1	2	1	1	1	1	1	1	1	1	1	1	1	1	1
		Total	-7	-7	-5	-4	-4	-4	-4	-6	-4	-4	-4	-4	-4	-4	-4	-4	-4	-4	6	-4	-4
Desintegração	Ruídos ou Vibração	Caráter (Ca)	-1	-1	-1	-1	-1	-1	-1	-1	-1	-1	-1	-1	-1	-1	-1	-1	-1	-1	1	-1	-1
		Importância (I)	1	1	1	1	1	1	1	1	1	1	1	1	1	1	1	1	1	1	1	1	1
		Cobertura (Co)	1	1	1	1	1	1	1	1	1	1	1	1	1	1	1	1	1	1	1	1	1
		Duração (D)	1	1	1	1	1	1	1	1	1	1	1	1	1	1	1	1	1	1	1	1	1
		Reversibilidade (R)	1	1	1	1	1	1	1	1	1	1	1	1	1	1	1	1	1	1	1	1	1
		Total	-4	-4	-4	-4	-4	-4	-4	-4	-4	-4	-4	-4	-4	-4	-4	-4	-4	-4	4	-4	-4
Laminação	Ruídos ou Vibração	Caráter (Ca)	-1	-1	0	0	-1	-1	-1	-1	-1	-1	-1	-1	-1	-1	-1	-1	-1	-1	1	-1	-1
		Importância (I)	1	1	1	1	1	1	1	1	1	1	1	1	1	1	1	1	1	1	1	1	1
		Cobertura (Co)	1	1	1	1	1	1	1	1	1	1	1	1	1	1	1	1	1	1	1	1	1
		Duração (D)	1	1	1	1	1	1	1	1	1	1	1	1	1	1	1	1	1	1	1	1	1
		Reversibilidade (R)	1	1	1	1	1	1	1	1	1	1	1	1	1	1	1	1	1	1	1	1	1
		Total	-4	-4	0	0	-4	-4	-4	-4	-4	-4	-4	-4	-4	-4	-4	-4	-4	-4	4	-4	-4
Extrusão	Ruídos ou Vibração	Caráter (Ca)	-1	-1	0	0	0	0	-1	-1	-1	-1	-1	-1	-1	-1	-1	0	-1	-1	1	-1	-1
		Importância (I)	1	1	1	1	1	1	1	1	1	1	1	1	1	1	1	1	1	1	1	1	1
		Cobertura (Co)	1	1	1	1	1	1	1	1	1	1	1	1	1	1	1	1	1	1	1	1	1
		Duração (D)	1	1	1	1	1	1	1	1	1	1	1	1	1	1	1	1	1	1	1	1	1
		Reversibilidade (R)	1	1	1	1	1	1	1	1	1	1	1	1	1	1	1	1	1	1	1	1	1
		Total	-4	-4	0	0	0	0	-4	-4	-4	-4	-4	-4	-4	-4	-4	0	-4	-4	4	-4	-4
Corte	Ruídos ou Vibração	Caráter (Ca)	-1	-1	0	0	-1	0	-1	-1	-1	-1	-1	-1	-1	-1	-1	0	-1	-1	1	-1	-1
		Importância (I)	1	1	1	1	1	1	1	1	1	1	1	1	1	1	1	1	1	1	1	1	1
		Cobertura (Co)	1	1	1	1	1	1	1	1	1	1	1	1	1	1	1	1	1	1	1	1	1
		Duração (D)	1	1	1	1	1	1	1	1	1	1	1	1	1	1	1	1	1	1	1	1	1
		Reversibilidade (R)	1	1	1	1	1	1	1	1	1	1	1	1	1	1	1	1	1	1	1	1	1
		Total	-4	-4	0	0	-4	0	-4	-4	-4	-4	-4	-4	-4	-4	-4	0	-4	-4	4	-4	-4
Mistura	Ruídos ou Vibração	Caráter (Ca)	-1	-1	0	0	0	-1	-1	-1	-1	-1	-1	-1	-1	-1	-1	-1	-1	-1	1	-1	-1
		Importância (I)	1	1	1	1	1	1	1	1	1	1	1	1	1	1	1	1	1	1	1	1	1
		Cobertura (Co)	1	1	1	1	1	1	1	1	1	1	1	1	1	1	1	1	1	1	1	1	1
		Duração (D)	1	1	1	1	1	1	1	1	1	1	1	1	1	1	1	1	1	1	1	1	1
		Reversibilidade (R)	1	1	1	1	1	1	1	1	1	1	1	1	1	1	1	1	1	1	1	1	1
		Total	-4	-4	0	0	0	-4	-4	-4	-4	-4	-4	-4	-4	-4	-4	1	-4	-4	4	-4	-4
Queima	Emissão de gases residuais	Caráter (Ca)	-1	-1	-1	-1	-1	-1	0	-1	-1	-1	-1	-1	-1	-1	-1	-1	-1	-1	1	-1	-1
		Importância (I)	2	2	1	1	2	1	1	2	2	1	1	1	1	1	1	1	1	1	1	1	1
		Cobertura (Co)	2	1	1	1	2	1	1	1	1	1	1	1	1	1	1	1	1	1	1	1	1
		Duração (D)	2	2	1	1	1	1	1	1	1	1	1	1	1	1	1	1	1	1	1	1	1
		Reversibilidade (R)	2	2	1	1	1	1	1	1	1	1	1	1	1	1	1	1	1	1	1	1	1
		Total	-8	-7	-4	-4	-6	-4	0	-5	-5	-4	-4	-4	-4	-4	-4	-4	-4	-4	4	-4	-4
Queima	Ruídos ou Vibração	Caráter (Ca)	-1	-1	-1	-1	-1	-1	-1	-1	-1	-1	-1	-1	-1	-1	-1	-1	-1	-1	1	-1	-1
		Importância (I)	1	1	1	1	1	1	1	1	1	1	1	1	1	1	1	1	1	2	1	1	1
		Cobertura (Co)	1	1	1	1	2	1	1	1	1	1	1	1	1	1	1	1	1	1	1	1	1
		Duração (D)	1	1	1	1	2	1	1	1	1	1	1	1	1	1	1	1	1	1	1	1	1
		Reversibilidade (R)	1	1	1	1	2	1	1	1	1	1	1	1	1	1	1	1	1	2	1	1	1
		Total	-4	-4	-4	-4	-7	-4	-4	-4	-4	-4	-4	-4	-4	-4	-4	-4	-4	-6	4	-4	-4
Estocagem	Armazenamento de produtos	Caráter (Ca)	-1	-1	-1	-1	-1	-1	-1	-1	-1	-1	-1	-1	-1	-1	-1	-1	-1	-1	1	-1	-1
		Importância (I)	1	1	1	1	1	1	1	1	1	1	1	1	1	1	1	1	1	1	2	1	1
		Cobertura (Co)	1	1	1	1	1	1	1	1	1	1	1	1	1	1	1	1	1	1	1	1	1
		Duração (D)	1	1	1	1	1	1	1	1	1	1	1	1	1	1	1	1	1	1	1	1	1
		Reversibilidade (R)	1	1	1	1	1	1	1	1	1	1	1	1	1	1	1	1	1	1	1	1	1
		Total	-4	-4	-4	-4	-4	-4	-4	-4	-4	-4	-4	-4	-4	-4	-4	-4	-4	-4	5	-4	-4

Weights assigned to valorise impact

Valuing impact			
Character (Ca)	Positive (1)	Neutral (0)	Negative (-1)
Importance (I)	High (3)	Average (2)	Low (1)
Coverage (Co)	Regional (3)	Location (2)	Puntual (1)
Duration (D)	Permanent (3)	Media (2)	Short (1)
Reversibility (R)	Irreversible (3)	Partial (2)	Reversible (1)

Table 1 - Valuing Impacts

Total Impact (IT): H= Ca. (I + Co + D + R) (1)

ANNEX

FEDERAL PUBLIC SERVICE REGIONAL COUNCIL OF ENGINEERING AND AGRONOMY OF RIO GRANDE DO NORTE CHECK-LIST FOR INSPECTING MINING COMPANIES - CREA -RN

INSPECTED BY:
DATE OF INSPECTION:
COMPANY NAME:

General company information

	YES	NO	REMARKS
01- Does the company have a mineral research application?			
02 - Is the company registered with CREA?			
03 - Does the company have a professional in charge of mine operations?			
04 - Is the company's professional registered with CREA?			
05 - Has the company registered its RAL (Annual Mining Report) with the DNPM?			
06 - Is the company registered with any environmental organisation?			
07 - Are the employees from the company itself (mine)?			
08 - Is the mine open-cast or underground?			
09 - Does the company produce industrial minerals that directly supply the construction industry, such as sand, gravel and ornamental stone (granite and marble)?			
10 - Were the management (owners of the company) kind to the inspection?			

Data relating to the company's geological information

	YES	NO	REMARKS
11 - Does the company have its own quarry for extracting raw materials?			
12 - Does the company buy raw materials from third parties?			
13 - Does the mine have a geological survey?			
14 - Does the mine have a topographer?			

15 - Is the ore extracted from the mine a metallic mineral?			
16 - Is the ore extracted from the mine a clay mineral?			
17 - Is the mine working on stable rock?			

Data relating to the company's rock dismantling stage

	YES	**NO**	**REMARKS**
18 - Is the mine extracted manually?			
19 - Does mining use mechanical equipment?			
20 - Does the mine work with explosives?			
21 - Is the explosive stored in a company warehouse?			
22 - Are explosives purchased by third-party companies?			
23 - Is the extraction equipment (loaders and lorries) in-house?			
24 - Do the benches have a lot of undergrowth?			
25 - Does the mine have its own accessories warehouse?			

Data on the company's processing stage

	YES	**NO**	**REMARKS**
26 - Does the company have a shed for equipment maintenance?			
27 - Does the company have a mineral processing department?			
28 - Does ore processing have crushers, mills and screens?			
29 - Does processing involve concentration stages such as flotation, magnetic and electrostatic separation?			
30 - Does the processing plant use gravimetric equipment such as simple chutes, oscillating tables, jigs and others?			
31 - Does the processing plant have a control room for its operations (video monitoring)?			
32 - Is the processing equipment modern enough?			
33 - Does the company have a system for processing raw materials such as metallurgy, ceramics, pottery and charcoal?			
34 - Does the company have a transport system for the ore produced (trains, conveyor belts, pipelines, lorries, etc)?			
35 - Is the equipment constantly inspected?			
36 - If the company works with flotation, what reagents are used?			

Data on occupational safety and environmental issues adopted by the company

	YES	**NO**	**REMARKS**

37 - Does the mine have any occupational safety signs?			
38 - Do the company's employees use PPE?			
39 - Is the mine environment unhealthy in any way, such as dust or high levels of noise?			
40 - If the mine is underground, does it have two alternative accesses?			
41 - Does the underground mine have adequate ventilation and lighting systems?			
42 - Does the underground mine offer the minimum safety conditions for workers?			
43 - Does the mine have an environmental recovery plan (PRAD)?			
44 - Does the mine have a water truck wetting down the access roads?			
45 - Does the mine have an area for disposing of the tailings discarded by the mine?			
46 - Does the company have any tailings dams?			
47 - Does the mine cultivate plant species native to the region?			
48 - Does the mine have a lake for reusing water?			
49 - Is the mine located near housing estates?			
50- Do machine operators receive specialised training before starting work?			

FINAL CONSIDERATIONS

In view of the above, I declare that the information collected is true.

Yours sincerely:

Christmas of 2015.

Printed by Books on Demand GmbH, Norderstedt / Germany